Temporālis
And Valid Reasoning?

By

Damon Dion Reed

Contents Table

Chapter 1: Earth 3

Chapter 2: Atomic Feedback 4

Chapter 3: Brain Bomb 2 6

Chapter 4: Mass Dependent Properties 10

Chapter 5: Protein Magnetospheres 14

Chapter 6: Catalysis Obsession 20

Chapter 7: Plasma Modulating Groups 23

Chapter 8: My Manhattan Project 28

Chapter 9: Boobs & Fusion 32

Chapter 10: Auto-Chiralization 36

Chapter 11: Catalytic Cells 38

Chapter 12: Histones 40

Chapter 13: MSSFs 43

Chapter 14: Krebs' Pickle 49

Chapter 15: Structure Equals Function 53

Chapter 16: Rates 59

Chapter 17: PiHKAL with THC and Nicotine 61

Chapter 18: Diels Alder 65

Chapter 19: Soooo? 69

References 70

Chapter 1: Earth

It was once thought that the Earth was the center of the universe, a waystation to nothing, butt data eventually gave way to science. And in as much as we've taken to crude bathroom humor, mostly because it doesn't make our heads hurt, there's still data patterns to be discovered in this branch of the universe. We just need to <u>know</u> where to look. And that is why I continually theorize. I'm like a cartoonish pointer dog that's rigid with determination: "Look people, patterns!" In any event, let us wonder. Let us muse. Let us wake every morning knowing there are repetitive patterns in the universe…even though we're not the center of it.

For those of you who don't know me, I like to get the point, leave it at that, and then talk shit about it later. Butt, I also have a playback, rerun, and/or echo. And based upon how I think the mind works, I have a sneaking suspicion that most writers have this talent. It's what makes us playfully dance with an idea until it makes sense. In any event, this is just another theoretical book, a literary echo if you may, to point researchers in the direction of patterns…in hopes that it will make the world a better place or "kill all humans"[2], whichever is easier.

Chapter 2: Atomic Feedback

If you believe humanity is the result of variance over multiple generations, then you should have no problem understanding atomic feedback. Quite simply, unique negative electronic environments force valence atomic orbitals into certain positions, which forces electrons to traverse certain pathways through or around the atomic nucleus. And as a result of this, the movement of negative Neuprotrons are modulated, which causes dissonance between Neuproz particles and eventually, the degradation of Neuproz particles.

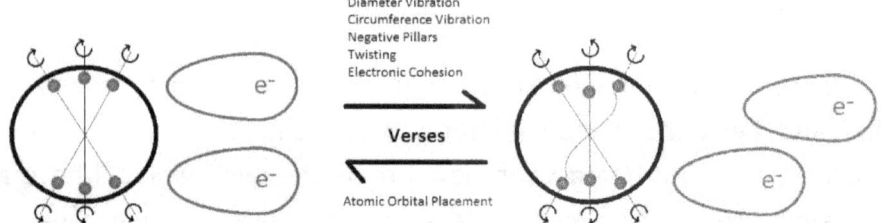

Figure 1: Energetic-feedback.

As you can see in this figure, several Neuproz Cluster factors influence the movement of electrons and subsequently their placement. But what the figure has difficulty portraying is that the "forced" positioning of the atomic orbitals has an influence on all those factors, which results in electronic dissonance within the Neuproz Cluster. And when the electronic dissonance is prolonged or intense, it results in Neuproz Cluster degradation, which will modify all the atomic orbitals or cause isotopic degradation. All of which, leads to a very interesting realization: **MOST** Neuproz Cluster degradations are UN-detectable.

Everybody with a brain and two cellphones to rub together, has been briefed on radiation, i.e. charged electronic entities that burst forth from atoms when diplomacy breaks down. But, in as much as we exist in a Quantum Dynamic universe, which is always degrading, AS WELL AS the fact that we exist in a negative branch of the universe and Neuproz Clusters contain TONS of positive energy, a vast majority of nuclear degradations never see the light of day...or notation within Nature articles. Maybe this is due to the negative environments we manipulate to detect energy in this negative branch of the universe or maybe it is simple because most nuclear degradations are **positive** and break down in the surrounding <u>negative</u> environment. All of which, makes me wonder: If elemental-half-lives are simply based upon DETECTABLE energy emission, then Houston...we have a problem?

In conclusion, the biofeedback system of dissonant atomic orbitals will cause dissonance within the Neuproz Cluster, which will eventually lead to degradation/release of energy. But, since the Neuproz Cluster is an overwhelmingly positive entity, most of the degradations are positive, which reacts with the negative plasma to render the "nuclear radiation" **undetectable**. Or simpler terms, there are millions of nuclear degradations, i.e. emissions, between each step in the isotopic degradation pathway, but since these emissions are mostly positive and react with the negative environment, they are NEVER detected. Or in the simplest terms, nobody notices your breath unless it smells. Or in scientific terms, when someone finally imagines a way to isolate stereo-isotopes, it will probably be another three-million years until someone imagines a way to detect EVERY electronic emission from the Neuproz Cluster...because they're mostly positive and immediately degrade in this negative branch of the universe.

Chapter 3: Brain Bomb 2

In the last chapter, I imparted to your neurons the concept of Electronic Isotopes…thank you very much and you're welcome. But, I lied about the timeframe with regards to someone imagining a method to isolate unique Electronic Isotopes, i.e. I did it last night. (Maybe I meant it in dog years since I'm a cartoonish pointer dog.) Any who, let's go back and peruse one step past Figure 1, i.e. Figure 2.

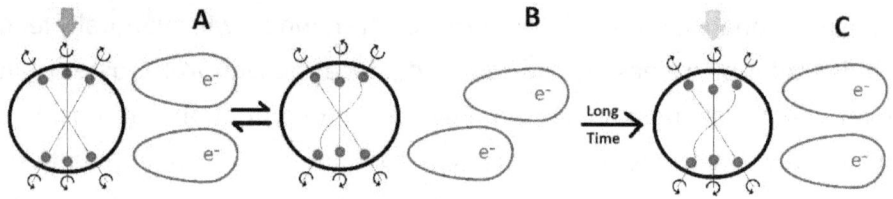

Figure 2: Eva Lovia long time distortion.

Now, to the normal eye, after being fucked for a 'Long Time', it appears that the "B" atomic orbitals have returned to their "A" position in the "C" part of Figure 2. Butt, you and I both know, the difference between the "A" & "C" Neuproz Clusters, notated by the red and orange arrows, means that the atomic orbitals are **different**. All of which, brings us back to those positive UN-detectable Electronic Farts, which degrade when they hit negative plasma.

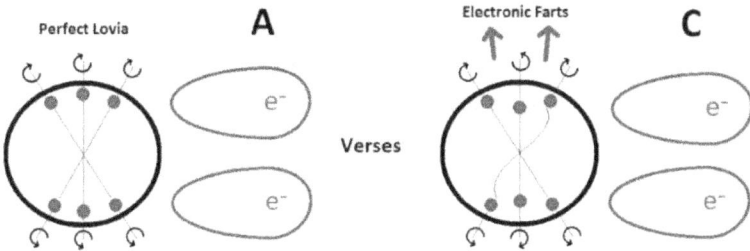

Figure 3: Farts be about?

Even though the dissonant "C" Neuproz Cluster has a different atomic orbital arrangement, the Electronic Farts increase the atomic orbital variance. Or in simpler terms, atomic orbitals are distorted by the plasma created by these Electronic Farts. All of which, sets the stage to detect subtle atomic orbital differences, i.e. Electronic Isotopes.

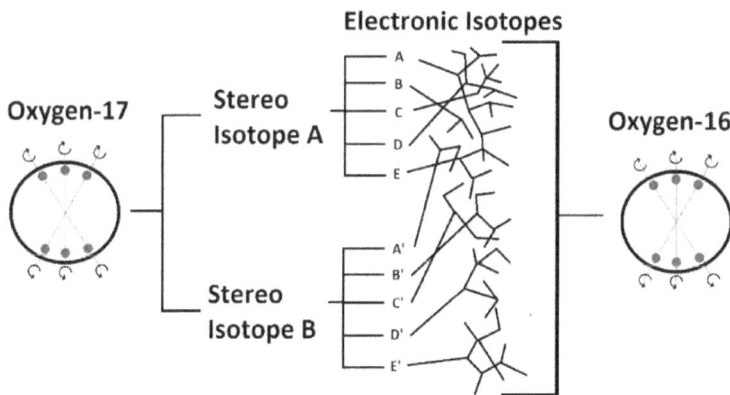

Figure 4: Complex Electronic Isotopes.

As you can surmise from this figure, each Stereo Isotope can have TONS of different Electronic Isotopes, which can vary based upon the Electronic Farts. All of which, **miraculously** results in the formation of Oxygen-16. (BTW, the electronic environment plays a HUGE part in the formation of UNIQUE Electronic Isotopes via atomic feedback, i.e. Chapter 2.)

Now, for those of you that didn't read the last chapter, the big Brain Bomb was the existence of Electronic Isotopes. (The minor Brain Poof was that most of the nuclear radiation isn't detected because it's positive and degrades in our negative branch of the universe.) All of which, when pieced together with the knowledge that atomic orbital placement is a function of plasma flow about the atomic nucleus, gives a METHOD to detect Electronic Isotopes.

Quite simply, each Electronic Isotope will have a unique electronegativity. Or in scientific terms, the arrangement of the **atomic orbitals**, i.e. electronegativity, is a function of the Neuproz Cluster and Electronic Farts. All of which means, each Electronic Isotope will RETAIN a slightly different amount of negative thermal energetic quanta. And as such, each Electronic Isotope will have a slightly different charge.

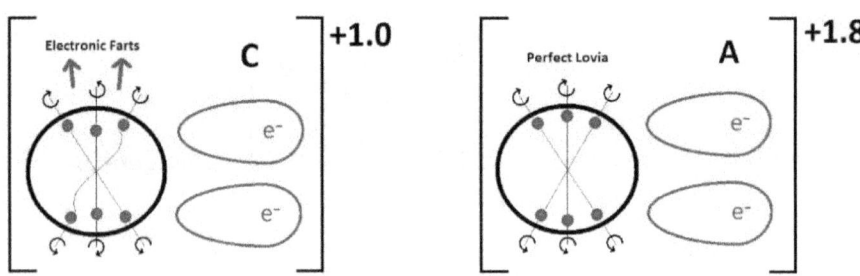

Figure 5: Atomic Orbital Placement verses Positivity.

Now I realize that the atomic orbitals look exactly the same in both Electronic Isotopes, but the miniscule differences in atomic orbital shape AND placement has made Electronic Isotope "C" more electronegative. Or in other words, Electronic Isotope "A" is LESS electronegative because it allows LESS plasma into its atomic orbitals and crevasses between the atomic orbitals, which makes it MORE positive. All of which, can be used to separate these unique Electronic Isotopes.

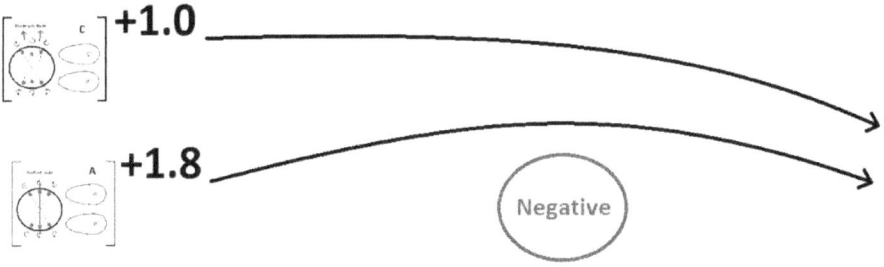

Figure 6: Electronic-isotope surfing.

If you make the Electronic Isotopes weave back and forth past tiny cold negative points, then their variance in positivity will result in different trajectories, which will allow the Electronic Isotopes to be separated. (BTW, Electronic Isotopes with greater electronegativities are more magnetically responsive, which might be helpful if the idea in Figure 6 doesn't work. Unfortunately, I won't talk about that until much-much later, i.e. a different book.) All of which, leads to the last question of this chapter: At what temperature do you store Electronic Isotopes such that they don't degrade into different Electronic Isotopes?

In conclusion, based upon Quanta Dynamics, it is conceivable that unique circumstances will render unique Electronic Isotopes with different electronegativities, which will allow for their variable refraction past multiple cold negative points to provide relatively pure Electronic Isotopes...Yeah! (Wait, what good are pure Electronic Isotopes?...Damn it, another question.)

Chapter 4: Mass Dependent Properties

The major problem of "information" being distributed via "text" is that people often miss the sarcasm. Granted, I often miss sarcasm, mostly because people aren't good it, but that is beside the point. The point is, Figure 4's conclusion was tots sarcastic. I mean, I even included **miraculously** to hint at the sarcasm. Or in simpler terms, the differential decay of hundreds of different oxygen-17 Electronic Isotopes will NOT forge ONE oxygen-16 atom. Now, with that aside aside, let's move onto something a bit more interesting: Mass Dependent Properties. Butt first, let's investigate Mass's evil-twin stepsister: Volume.

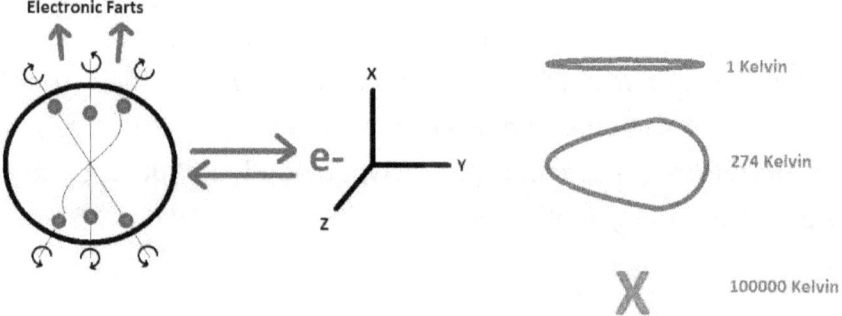

Figure 7: Atomic Orbital behavior.

Atomic volume is a function of temperature, which is related to mass. And as you can see in this figure, the atomic orbital volume is determined by electron movement in the X,Y, and Z planes. Or in simpler terms, when there isn't much heat, e.g. 1° Kelvin, the atomic orbital is nothing more than a "spring type" movement away from the positive atomic nucleus, which is modulated **slightly** by the movement of positive charges within the atomic nucleus. Next, in the presence of some heat, e.g. 274° Kelvin, the atomic orbital looks like an atomic orbital. And finally, in the presence of **insane** amounts of heat, e.g. 100,000° Kelvin, there is NO atomic orbital because electrons can NOT create a repetitive path, i.e. atomic orbital. All of which, brings us to some intuitive logic. (Honestly, I have no clue when to use an e.g. instead of an i.e., so I just toss them in randomly.)

For years, scientist have known there is a temperature correlation to the oxidation of atoms, but they have refused to succumb to this logic: If heat causes the release of a NEGATIVE electron, then HEAT must be negative. Now, the reason why I bring this up is: There are scientists that claim they have created "almost" absolute zero environments, but they haven't detected any radiation. Or in simpler terms, look down.

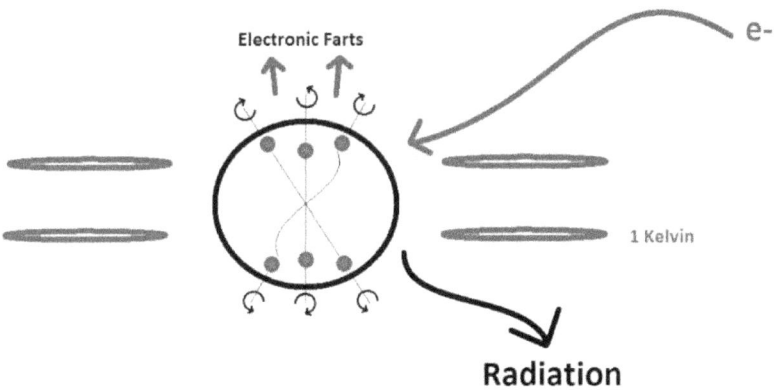

Figure 8: Close to ZERO radiation.

Quite simply, variable reduction-oxidation reactions in this "almost" absolute zero environment will send electrons hurling towards the atomic nucleus, which will result in a "fair" amount of radiation. (BTW, by "fair" amount, I mean an EXTRA electron will result hell-n-brimstone, i.e. X-rays, gamma rays, and possibly Ray Charles.) Or in simpler terms, scientists are still really-really far away from absolute zero...relatively speaking of course. All of which, brings us back to Mass and its evil-twin stepsister, Volume.

For those of you who haven't been exposed to radiation or the concept of Avogadro's number, you're lucky. Quite simply, Avogadro's number correlates MASS to the number of atoms. But, since MASS is a temperature dependent property, i.e. negative heat determines the mass, Avogadro's number is more interesting than previously thought. Or in complex terms, Avogadro's number is a harmonic within this negative branch of the universe, NOT just random number. Or in postulative terms, if you add Avogadro's number to the MASS of a thermal energetic quanta, then you'll get the MASS of an electron. Or in simpler terms, I propose that the mass of a thermal energetic quanta is about 3.088×10^{-54} kg. As for the reason why this difference harmonically re-appears about mass as it relates to the number of atoms is: Thermal energetic quanta are major FACTORS in determining the mass/volume of matter in this negative branch of the universe. All of which, brings us back to Electronic Isotopes.

The first thing scientists will do after isolating Electronic Isotopes is have a conference. And at this conference, they'll spend most of their time arguing over which Electronic Isotope has more energy. All of which, will ultimately be focused back to the electronegativity of the Electronic Isotope. Therefore, let me save those scientists some time and set the stage for them.

First, scientists should determine the number of thermal energetic quanta in each Electronic Isotope at a given temperature. Next, identify their electronegativities. And finally, shoot the Electronic Isotopes at each other and see how they explode. (Oops, sometimes I slip back into the OLD way of thinking…Take Two:) And finally, if a pattern can't be identify, just flip a coin and get on with their lives.

In conclusion, in as much as Avogadro's number is MASS dependent property, it shouldn't be a surprise that it's simply a harmonic within the "mass equation"…in this negative branch of the universe. Or in other words, I propose that a single Thermal Energetic Quanta weights about 3.088×10^{-54} kg. (If my calculations are right.) Or in converse terms, the reason why Avogadro's number exists is because an electron is 6.022×10^{23} BIGGER than a thermal energetic quanta. And finally, you're NOT anywhere close to ABSOLUTE zero UNLESS you're detecting a large amounts of radiation. (FYI, Madame Curie feels your pain if you've been trying to create an absolute zero environment without the appropriate radiation protection...not really, she's dead.)

Chapter 5: Protein Magnetospheres

It's funny, Styrofoam is a plastic, e.g. polymer, but it has AMAZING properties in comparison to other plastics. All of which, is somehow correlated to the previous chapter and proteins...hopefully. In any event, Styrofoam's polymer is different because it has an aromatic ring on each monomer. Or in simpler terms, look down.

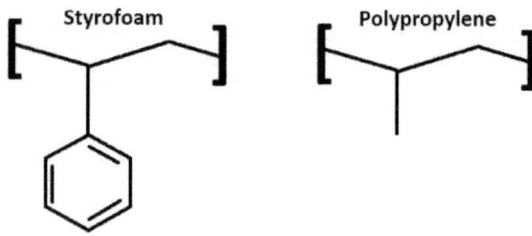

Figure 9: Aromatic Dingleberry.

As you can see, Styrofoam has this HUGE aromatic ring dangling off, which is the REASON why it is a wonderful insulator. Unfortunately, to make you believe this, I'll have to label the OLD definition of insulation as schizophrenic: Thermoses insulate better when they have a vacuum chamber, i.e. no air; Polystyrene insulates because it has tons of trapped air. Hopefully, you see the dichotomy in the OLD logic. In any event, let's dismiss this absurdity and investigate a more logical explanation.

This might be surprising to you, but it took a couple hundred brilliant scientists several years to identify the concept of aromaticity. And oddly enough, the dude that figured it out, said he discovered it in a dream.[3] Fortunately, with the advent of Quanta Dynamics, i.e. the expulsion of "random" quantum tunneling, it's time to give aromaticity make-over! (OMG...it's going to be so beautiful when I get done with it! #MakeOver)

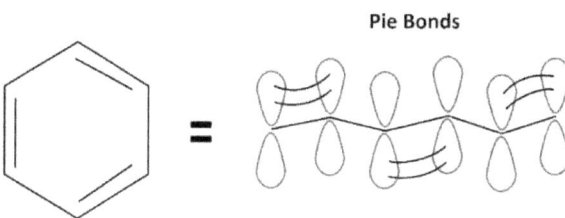

Figure 10: Just like grandma used to make?

I don't know why they call them pie bonds, but I think it has something to do with benzene looking like a pie. In any event, these pie bonds move around the benzene ring because they're all conjugated. All of which, creates a wave-looking thing.

Figure 11: Pie bonds.

In olden times, i.e. last week, this wave-looking structure was just a quant observation, which resulted in a chuckle from time to time...probably at a conference somewhere. But, in terms of Quanta Dynamics and electrons recharging by searching for the positive charges within the atomic nucleus, it is a LOGICAL pattern to electron movement. All of which, brings us back to Quanta Dynamic's basis of atomic orbitals: The protection of the positive charges within the Neuproz Cluster. Or in simpler terms, atomic orbitals push back the insanely rude and intrusive negativity in this negative branch of the universe. Or in the simplest terms, look down.

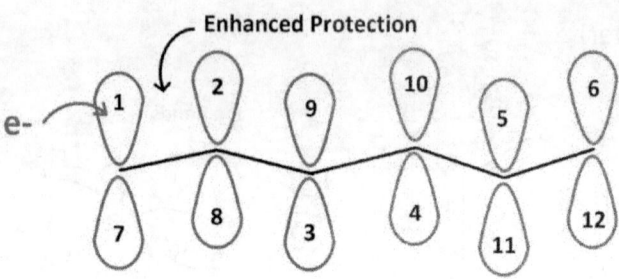

Figure 12: Vroom Vroom though the numbers!

The PRIMARY basis of atomic orbitals is to protect the positive Neuproz Cluster. Therefore, as the electrons REPEATEDLY surge through ALL the 'LINKED' numerical atomic orbitals, their movement between the atomic orbitals, i.e. the "Enhanced Protection" zone in Figure 12, will protect the positive Neuproz Cluster from negative plasma. Or in Quanta Dynamic terms, electrons in aromatic pie bonds move at a faster harmonic than electrons in normal valence atomic orbitals. (FYI that will become clearer in subsequent books.)

Now, for any person with a brain larger than a pea, this is more than remotely disturbing. For starters, this results in chirality when an aromatic ring is not free to rotate. Or in simpler terms, the movement of the electrons about the pie bonds, either clockwise or counterclockwise, when the aromatic ring isn't able to rotate, will result in chirality. (FYI, this is strangely amusing when it comes to people trying argue in favor of the OLD theories, i.e. there are several examples of chiral aromatic rings in most undergraduate chemistry textbooks.) In any event, let's take a moment and compare/contrast aromatic rings and super-conducting magnets.

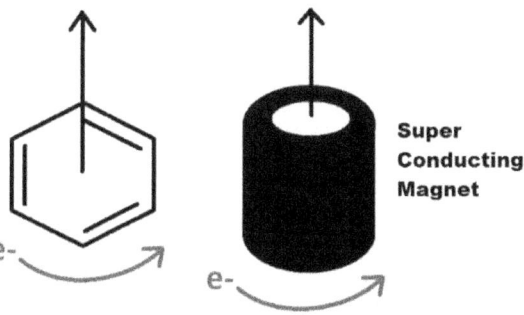

Figure 13: Seeing Double?

As you can see, a super conducting magnet has electrons surging around in a circle and the aromatic ring has electrons surging around in a circle. Also, the super conducting magnet modulates the Earth's magnetosphere to create a strong magnetic moment and the aromatic ring modulates the Earth's magnetosphere to create a strong magnetic moment...relatively speaking of course. And finally, a super conducting magnet accomplishes this with **billions** of electrons, but the aromatic ring accomplishes this with only **six** electrons. (Spoiler Alert, this chances to 18 electrons in about six paragraphs.)

With all that in mind, the insulating behavior of Styrofoam is a FUNCTION of the magnetically active aromatic rings randomizing the plasma movement. Or in water terms, Styrofoam's aromatic rings are like waters' positive protons, which inhibit the diffusion of negative thermal energetic quanta. Granted, protons are positive and aromatic rings are negative, but they both randomize the movement of negative plasma. All of which, finally, brings us to protein magnetospheres.

Temperature is based upon plasma and the flow of plasma is based upon the order of atoms...or disorder. Therefore, the flow of plasma about different proteins, i.e. highly ordered amino acids, determines the residual plasma within the protein complex. Or in simpler terms, each protein has a unique thermal capacity, which is a function of the plasma movement within the protein. (FYI, it is important to remember that the placement of water within proteins is also a major factor in determining its unique thermal capacity.)

Butt, having said all that, I would have to give-up my pedantic-ID if I didn't bring one more thing to your attention: The harmonic aromatic wave pattern is a function of the environment. Or in simpler terms, look down...as if you needed to be reminded that this is how English books function.

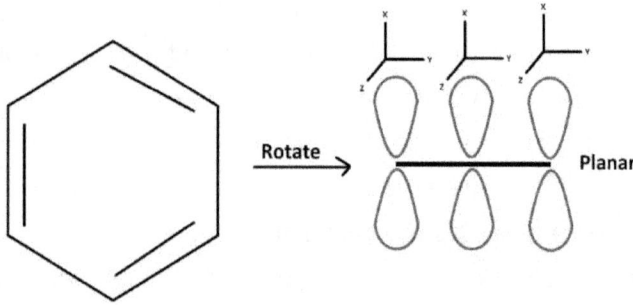

Figure 14: For Organic Chemistry haters!

As you can surmise from this figure, when you rotate the benzene ring, the atomic orbitals will be "perpendicular" to the aromatic ring. Now, the reason why "perpendicular" is in parenthesis is because the world is a DYNAMIC place. Therefore, any angular movement of ANY atomic orbital will "tune" the aromatic ring, which I will talk about in later chapters. But, let me reassure you of one thing: The rate difference between negative Neuprotrons and electrons allows for multiple harmonic rhythms...even though most people can't hear or imagine them.

Unfortunately, after writing the last point, I was driving home when I realized something extremely stupid, but completely intuitive. Quite simply, the ONLY reason why aromatic pie bonds have such an extraordinary stability is because the Chorus Line. Now, I realize that this makes absolutely no sense right now, but it will make sense when you read my subsequent books...hopefully. (#SuperForeshadowing) In any event, let me just say that I have it on good authority that the sigma bonds connecting the carbons within benzene ALSO are rotating electrons around the aromatic ring. Or in simpler terms, the reason why an aromatic ring mimics a super conducting magnet is because the enhanced electron movement about the sigma bonds. (This means there are 18 total "aromatic" electrons in benzene.) Or in figurative terms, look down.

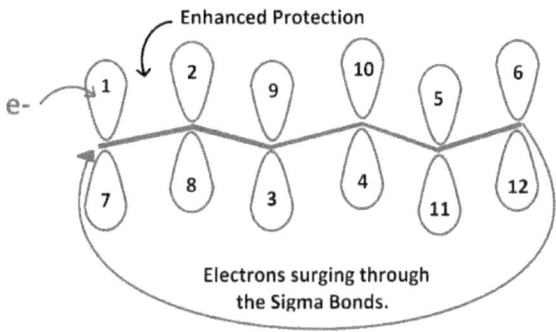

Figure 15: If you look really close, I chopped-off the "g" in Sigma.

As you can see, there is an increase in electron density about the sigma bonds as a result of the electrons fluxing around in a circle. (Trust me, I'll talk about this in a future book because it's very-VERY interesting.)

In conclusion, a lot of things are amazing. But, the ability of Quanta Dynamics to expand the understanding around aromaticity is pretty awesome. (#CirclePun) Or in other terms, because of the surging of the electrons about aromatic sigma and pie bonds, benzene mimics a super-conducting magnet. (Technically, super-conducting magnets mimic benzene rings.) All of which, is the reason why Styrofoam is such a good insulator, e.g. randomization of the plasma flow by **aromatic** super-conductors. And finally, it is pretty amazing how uniquely hydrated protein structures, which contain chiral aromatic rings, can maintain UNIQUE plasma levels...even when they're in a body with a completely different temperature.

Chapter 6: Catalysis Obsession

Having brought up the concept of aromatic entities as it relates to temperature and proteins, I feel slightly obligated to expand this into catalysis. Luckily, I have an example that supports the current definition of catalysis, but also broadens the definition to include Quanta Dynamics. I give you...Compound X, which was named by Professor Bergmeier because the NMR data was weird.[4] (FYI, the name stuck after getting the x-ray data because there is NO nomenclature for "catalytic" molecules like this.)

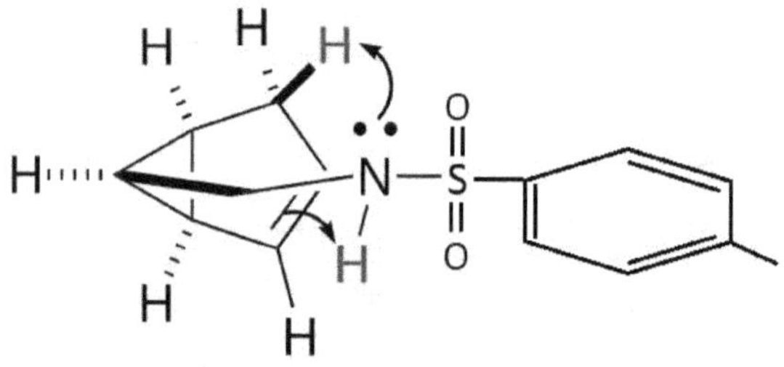

Figure 16: Compound X.[4]

After a bit of arguing, Professor Bergmeier and I finally agreed that this was probably the structure of Compound X. But, when I presented Professor Bergmeier with the previous research on the topic, things got even weirder. Quite simply, with one SIMPLE adjustment to the aromatic ring, i.e. the removal of the methyl group, pervious researchers had to use 100° NMR conditions to resolve the NMR data, i.e. create a NMR spectra with understandable splitting patterns.[5]

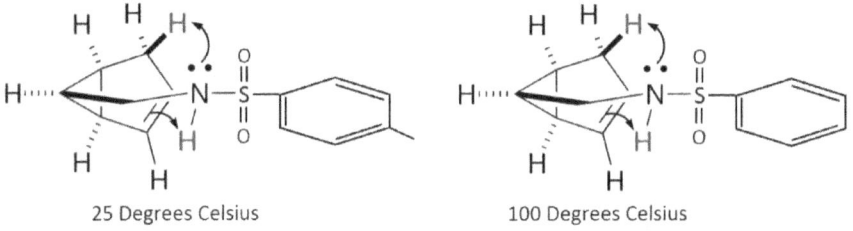

25 Degrees Celsius   100 Degrees Celsius

Figure 17: Heat Jiggle-Jiggle[5]

As you can surmise from this figure, the "catalytic" nature of the nitrogen seems to be linked to the methyl group and/or HEAT. Or in chemistry terms, aromatic modulation by the methyl allows the nitrogen to display catalytic properties at a LOWER temperature. Or in Quanta Dynamic terms, the harmonic modulation of the aromatic ring by the methyl results in an increase in temperature about the "catalytic" nitrogen. Or in simple and straight forward terms, the methyl group modulates the aromatic ring so that it focuses magnetism and plasma at the nitrogen to make it catalytic.

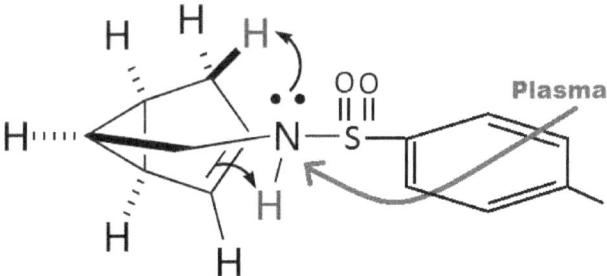

Figure 18: Focusing hard?

Hopefully, you are extremely perturb that I WILDLY jumped from aromatic rings being miniature super-conducting magnets to them being plasma pumps. Having said that though, I hope you are able to understand that a NEGATIVE magnetic breeze will also blow NEGATIVE plasma in the same direction. (#EquatorHeatZone)

With all this in mind, hopefully, a new understanding of amino acid secondary structures appears bright eyed and bushy tailed. For example, beta pleated sheets can be viewed as structural reinforcement to protein structures as well as a thermal insulators to underlying amino acids. As for alpha helixes, they can be structural reinforcement, an insulator, or possibly a tunnel-esk structure that focuses negative plasma onto certain amino-acids, which causes the protein to jiggle-jiggle. (FYI, the specific jiggle-jiggle of catalytic proteins allows the "catalytic center" to do its patented suck-n-spit thang.)

In conclusion, it has long been know that there is a temperature factor to catalysis. Or in Quanta Dynamic terms, a certain level of plasma is required to support vibrational-catalytic atomic orbital arrangements. Next, since aromatic rings are miniature super-conducting magnets, they also have the propensity to blow negative plasma, which results in a very interesting secondary function to secondary amino acid structures. All of which, results in Quanta Dynamics adjusting the theory about aromatic electron withdrawing and donating groups...albeit in the next chapter.

Chapter 7: Plasma Modulating Groups

So anyway, I tots need to write a chapter on the concept of electron donating and withdrawing groups because let's be honest, a lot of shit has changed since the initial contemplation of functional groups participating in aromatic reactivity. All of which, results in a heart-warming review of all the stuff you probably tried to forget.

^{1}H 1,2,3							Atomic Number → ^{15}P Isotopes										^{2}He 3,4
^{3}Li 6,7	^{4}Be 9											^{5}B 10,11	^{6}C 12,13,14	^{7}N 14,15	^{8}O 16,17,18	^{9}F 19	^{10}Ne 20,21,22
^{11}Na 23	^{12}Mg 24,25,26											^{13}Al 27	^{14}Si 28,29,30	^{15}P 31	^{16}S 32,33,34,36	^{17}Cl 35,37	^{18}Ar 36,38,40
^{19}K 39,40,41	^{20}Ca 40,42,43,44,46,48	^{21}Sc 45	^{22}Ti 46,47,48,49,50	^{23}V 50,51	^{24}Cr 50,52,53,54	^{25}Mn 55	^{26}Fe 54,56,57,58	^{27}Co 59	^{28}Ni 58,60,61,62,64	^{29}Cu 63,65	^{30}Zn 64,66,67,68,70	^{31}Ga 69,71	^{32}Ge 70,72,73,74,76	^{33}As 75	^{34}Se 74,76,77,78,80,82	^{35}Br 79,81	^{36}Kr 78,80,82,83,84,86
^{37}Rb 85,87	^{38}Sr 84,86,87,88	^{39}Y 89	^{40}Zr 90,91,92,94,96	^{41}Nb 93	^{42}Mo 92,94,95,96,97,98,100	^{43}Tc 98	^{44}Ru 96,98,99,100,101,102,104	^{45}Rh 103	^{46}Pd 102,104,105,106,108,110	^{47}Ag 107,109	^{48}Cd 106,108,110,111,112,113,114,116	^{49}In 113,115	^{50}Sn 112,114,115,116,117,118,119,120,122,124	^{51}Sb 121,123	^{52}Te 120,122,123,124,125,126,128,130	^{53}I 127	^{54}Xe 124,126,128,129,130,131,132,134,136
^{55}Cs 133	^{56}Ba 130,132,134,135,136,137,138	Lantha-nides	^{72}Hf 174,176,177,178,179,180,	^{73}Ta 180,181	^{74}W 180,182,183,184,186	^{75}Re 185,187	^{76}Os 184,186,187,188,189,190,192	^{77}Ir 191,193	^{78}Pt 190,192,194,195,196,198	^{79}Au 197	^{80}Hg 196,198,199,200,201,202,204	^{81}Tl 203,205	^{82}Pb 204,206,207,208,	^{83}Bi 209	^{84}Po 209	^{85}At 210	^{86}Rn 222
^{58}Ce 136,138,140,142	^{59}Pr 141	^{60}Nd 142,143,144,145,146,148,150	^{61}Pm 145	^{62}Sm 144,147,148,149,150,152,154	^{63}Eu 151,153	^{64}Gd 152,154,155,156,157,158,160	^{65}Tb 159	^{66}Dy 156,158,160,161,162,163,164	^{67}Ho 165	^{68}Er 162,164,166,167,168,170	^{69}Tm 169	^{70}Yb 168,170,171,172,173,174,176	^{71}Lu 175,176				

Figure 19: Spectrum Table...technically not a figure.[6]

On the **right**, we have **electronegative** atoms, i.e. the negative plasma suckers. On the <u>left</u>, we have the <u>electropositive</u> atoms, i.e. the non-suckers. (FYI, the noble gases could be on either the right or left side because they're freakishly somewhere between these two extremes, but that is beside the point.) The point is, negative plasma is allowed to flux closer to the positive atomic nucleus, which creates a slightly more negative atom. In any event, here is the spectrum:

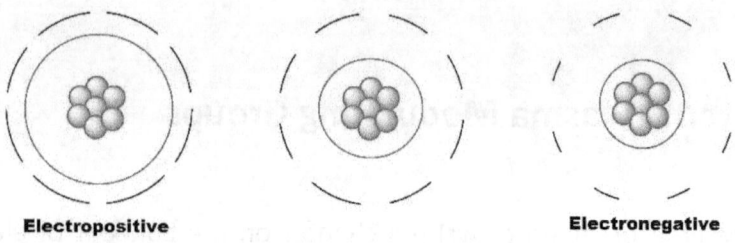

Figure 20: Electropositive verses Electronegative.

Now, as you can surmise, the porousness of electronegative atoms allows more negative plasma to surge closer the positive atomic nucleus, which causes degradation in the Neuproz Cluster and subsequently Isotopic Relaxation to create a spectrum of Isotopes in the periodic table. Or in Quanta Dynamic terms: Asymmetric placement of **extra neutrons** in electronegative atoms results in the asymmetric placement of the atomic orbitals; Asymmetric atomic orbitals result in large POCKETS between atomic orbitals, which allows more negative plasma to be contained about the atom; The closer the negative plasma is to the atomic nucleus, the easier plasma slippage can cause Isotopic Relaxation. (FYI, for those nerds that really have nothing better to do than critic my ideas, which is a huge honor, I need to point-out one peculiarity: Increased levels of PLASMA will cause Isotopic Relaxation of electronegative atoms, but increased levels of ELECTRONS will NOT have the same effect. Quite simply and logically, increased levels of ELECTRONS will result in the degradation of electropositive atoms into Noble Gases. As for the reason why, it's a momentum thing: Electrons got a shitload.)

Hopefully, your head doesn't hurt because here comes the fun part: Recalibrating your cognition of electro-donating verses electro-withdrawing groups in terms of Quanta Dynamics. And to aid in this recalibration, let's obsessively return to Compound X...from the previous chapter.

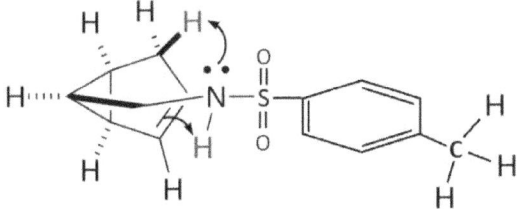

Figure 21: Compound X's little proton piggies.

To aid in your cognition, I've added the POSITIVE protons, i.e. instead of a simple line to signify the methyl group, I've added carbon and the H's on the right. NOW, let's review the __old__ theory of electron donating groups.

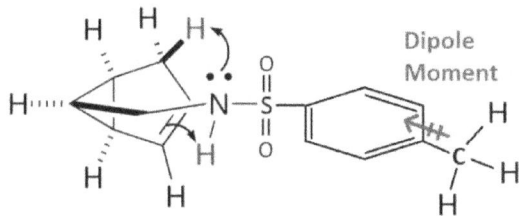

Figure 22: OLD Theory with a pointed cane.

The __old__ theory of electron donating groups is as follows: EVEN THOUGH CARBON FORMS FOUR EQUALLY SHARING TETRAHEDRAL CHEMICAL BONDS, somehow a bit of "ELECTRON density" surges towards the aromatic ring, which makes the aromatic ring more electron-rich. And now, here's Quanta Dynamic's view of electron donating groups, i.e. the QD-revision.

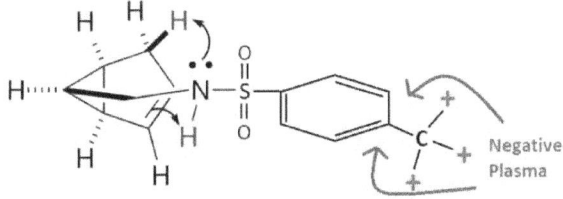

Figure 23: OMG, what happened to the protons?

As you can see, the QD-revision really isn't much of a revision. It's more of a SIMPLE truth hiding under misleading notation. Or figurative terms, those pesky NAKED positive protons suck negative plasma towards the aromatic ring, which is miniature super-conducting magnet. (FYI, **aromatic pie bonds** are HORRIBLE at corralling negative plasma, but the **aromatic ring** is great at focusing negative plasma. All of which, I'll explain in subsequent books.) In any event, let's use Quanta Dynamics to look at more examples...YEAH!

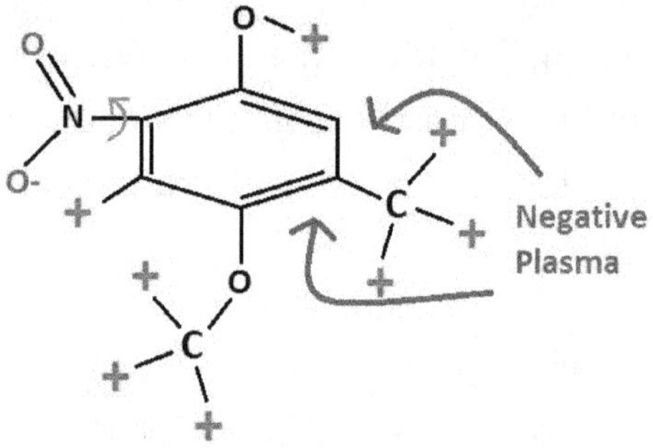

Figure 24: My God, repulsive naked bodies!

As you can see, simply noting the **charges** really does open-up your mind to the movement of negative plasma around an aromatic ring. Unfortunately, it's a little more complicated than looking at naked charges. Therefore, I'll leave it here for now. But, since this shit is so exciting (cough), let me share one more figure with you before I move onto the next chapter.

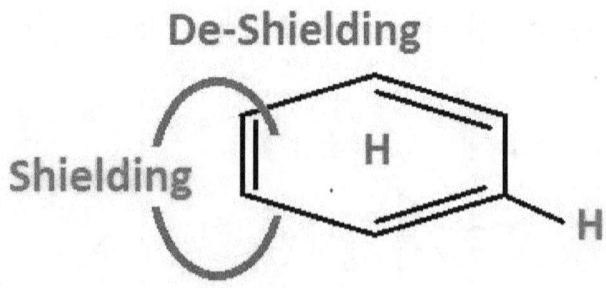

Figure 25: Finally, a bit of HUMAN decency!

In my subsequent books, I will dwell on this extremely ODD aromatic NMR pattern. But, if you're an overachiever, which I hope you are, then you can think ahead by smashing Figure 17, Figure 24, and all the previous stuff I've mentioned with regards to atomic orbitals and plasma.

In conclusion, even though I have only cracked the door with regards to revising electron donating and withdrawing groups in reference to aromatic systems, mostly by just notating the charges, it is delightful to know that there are nerds out there that will be mulling over this crap until I write another book. (FYI, it's not in my next book. It's in the book after that.) And finally, just so I don't hurt any brains, the De-Shielding effect of protons near the middle of the aromatic ring is caused by the flow of NEGATIVE plasma through the aromatic ring, which distorts the atomic orbital and SUCKS the POSITIVE proton away, i.e. shifts the proton in the NMR spectrum. (BTW, I always mess-up the up-field/down-field crap in NMR spectrums, which is why I didn't mention it.)

Chapter 8: My Manhattan Project

So where does one begin in theorizing a device that can kill large portions of society? It's simply, you need to start with DENIAL. First, you have to tell yourself that GOOD people will ALWAYS rise-up and protect the most venerable people...because man and woman are ultimately GOOD. Second, you lie to yourself and say that this technology will produce lots of clean energy and ultimately free humanity from Earth. And finally, you justify this device as a means to pussy and/or penis, whichever you desire. Therefore, with every level of emotional blowback covered, let's create a device that will obliterate large swaths of humanity, i.e. the water bomb.

The theory behind the device is relatively simple. First, heavier elements are only forged in dying stars and water is only forged from the decay of heavier elements in dying planets. Second, even though young protons are insanely stable as it relates to the thermal energy in young stars, old protons are NOT stable to deep-dark-cold positive space. Thirdly, and most importantly, doubly negative oxygen, in deep-dark-cold positive space, degrades and spews protons and neutrons everywhere, which is the method by which the energy from lighter elements is recycled and stored in heavier elements. But, before we get to the technical specifics, let's review how WWIV will probably begin.

TRUST...it is kind of important when two people are pointing guns at each other, which is a metaphor for the nuclear arms race. Unfortunately, humanity isn't very trusting. Also, how many nuclear weapons you have often determines how LARGE your part is in the negotiations process...to ANYTHING. All of which, suggests that WWIV will probably start as a result of an un-sanctioned nuclear program...or a nation that is based upon lies. Or in simpler terms, nuclear reactors continually cause the degradation of protons, which will result in doubly ionic oxygen **IF** heavy water is NOT continually added to the reactor pool. Or in the simplest terms, an un-sanctioned nuclear program will probably learn the hard way, i.e. big bomb, and then blame another country for the explosion, i.e. espionage. Then, everybody will pick sides based upon who they TRUST and wala...WWIV! (BTW, based upon how stupid people are with regards to religion, these TRUST-lines will probably be based upon religion.) In any event, back to the specifics of this wonderful device...for everyday living.

Let's say you have an old nuclear reactor with some wonderfully "cured" water molecules, i.e. water molecules with OLD protons. Now, at some point, when enough protons have gone boom, the doubly ionic oxygen will go boom. But how do you weaponized it? Well, you could invade a country, build them a nuclear reactor, and then let it produce energy for several years UNTIL it goes boom, but that's way too much of work. (#Diplomacy) Also, you're supporters might mistake your actions as altruistic instead of powerful. In any event, how do you weaponized "old" water? Well, how about we use a device for everyday living?

You might not know this, but it took a SERIOUS amount of advertising to get old-people to believe that microwaves didn't cause radiation. Maybe it was because of the cold-war or maybe, the sale of microwaves was simply an INGENIOUS marketing technique to keep people from freaking-out about nuclear war? (OMG, I think I'm onto something. First you introduce the microwave. Next you start a whisper campaign about how microwaves cause radiation.   Then you advertise microwaves until every home in America has one! And that ladies and gentlemen, is how GE ended the cold war, not Reagan. Hmm, does that relate to Iraq somehow? Or, are people still FREAKING-THE-FUCK-OUT about airplanes?) In any event, it's totally IRONIC because negative microwaves do cause radiation, i.e. the degradation of protons, by distorting water's atomic orbitals. (FYI, conventional ovens and microwaves BOTH cause protons to degrade.) Or in Manhattan Project terms, nuclear bombs are packed with explosives, which creates a lot of heat, like a conventional oven, and aids in the cascading uranium reaction...via atomic orbital distortion. Therefore, instead of using HUGE elements an tons of TNT, we're going to use a small element, e.g. water, and some microwaves.

With most of the theoretical and technical specifics out of the way, it's time to FREAK-THE-FUCK-OUT because water bombs won't be radioactive until after they go boom-boom. Or in other words, the decay of one or two oxygen atoms will only result in less basic water, which will be less explosive. (BTW, for those of you who don't know this, nuclear warheads ALSO become less destructive as they age, but nobody knows how much since everybody stopped testing nuclear bombs. All of which, is the reason why America and Russia don't want young countries developing nuclear weapons. Wow, the geopolitical ramifications of this chapter are kind of haunting.) In any event, we've already put in place blowback precautions, so let's continue forward.

After obtaining cured water from a nuclear reactor, you just need to identify the correct amount of microwave radiation to create a super-critical mass of double ionic water, which might be trickier than you think. Or in simpler terms, nobody has figured out how to determine the age of protons...since we ALL believe in the second law of thermodynamics, i.e. matter can neither be created or destroyed. Or in IDIOT POLITICAL terms, a scientist won't be able to get funding to study a method to find the relative age of protons because most of our POLITICIANS are idiots. Therefore, without the research to identify the relative age of protons, the USA's water bombs could go boom while they're in a plane or simply squish against the ground without going boom. (Based upon the fact that most politicians don't believe in science OR DATA, I have it on good faith WWIV will probably be the result of HORRIBLE data sets that will cause one of America's water bombs to go boom on American soil, which America will blame on another country because AMERICA DOESN'T MAKE MISTAKES!)

In conclusion, TRUST is a big thing and people always freak-out. (I'm still trying to figure-out how to put an airplane in EVERY American house, like a microwave, so that people stop FREAKING-THE-FUCK-OUT! And you'all thought my idea of flying cars was crazy…ha!) But more importantly, based upon Quanta Dynamics, a critical mass of double ionic oxygen will have MORE destructive power per mass because 99% of the SMALL oxygen atoms will degrade and expand in comparison to the expansion of only about 35% of the uranium's critical mass. Or in more specific terms, 75% of the uranium within the critical mass ONLY degrades by releasing a few protons, which results in the radioactive-fallout. Or in the simplest terms, water bombs will have an insane amount of expansion-power with very little amount of radiation, which will allow the USA to kill even more innocent people…(cough)…I mean, the United States DOES NOT MAKE MISTAKES! #GodBlessAmerica (BTW, I've got a much better way to make a water bomb, but I'm not going to tell you'all because I don't **TRUST** you.)

Chapter 9: Boobs & Fusion

I realize there are several opinions about the idea of fusion releasing energy. But in as much as fusion formally releases a neutrino and negative Neuprotron, I figured you'all would be down with a semantic argument. Unfortunately, as always, I overestimate the rationality and reasoning of other humans. Therefore, let me try and explain the logic of HOW fusion releases energy…boobs edition.

For those of you who don't know this, the US military owns devices that are formally known as 'fusion' devices, i.e. hydrogen bombs…I think. In any event, they're called 'fusion' devices because the energy RELEASED by these devices CANNOT be **accounted for under quantum mechanics**, which is because quantum mechanics is based upon the second law of thermodynamics. Or in simpler terms, if quantum mechanics can NOT explain all the energy released, THEN fusion must release energy and stars must be releasing energy as a result of fusion. (#WeirdLogic) Butt, what scientists are forgetting about is: BOOBS.

As a man living in a semi-enlightened society, I no longer look up at the stars and wonder what humanity might accomplish someday. What I do do, is look at boobs: Big boobs, little boobs, and every type of boob in between.   There're all magnificent as long as they come in pairs…single boobs are unnerving because they remind me of my own mortality. (FYI, single boobs and mortality is a perfect metaphor for Quanta Dynamics. Also, I also like butts.) In any event, before I venture any further into the realm of perfect boobs, I must make a weird observation about our sun, which shines down upon all the beautiful boobs…jiggling, wiggling, and bouncing up and down.

Our sun, for as old and yellow as it is, contains hydrogen and helium. Actually, let me re-clarify that statement: Scientists **think** our sun has helium and hydrogen because our sun has similar line spectrums to hydrogen and helium. Now, for those of you who haven't had the pleasure of taking freshman physics, line spectrums are kind of funny. Quite simply, under certain conditions, which includes a crap-ton of electrons, hydrogen and helium release specific photon spectrums. Or in Quanta Dynamic terms, the electrons of hydrogen and helium repetitively crash and degrade to release SPECIFIC photons under SPECIFIC conditions…imagine that? In any event, not to blow up the complete basis of astrophysics' elemental determination of stars, but our sun puts out a crap-ton of energy, which has some SPECIFIC photon concentrations.   Or in the simplest terms, there is ONLY **ONE** explanation for how the sun's NUCLEAR environment releases the same electron degradations as the electron degradations on Earth, i.e. the sun has hydrogen and helium. On second thought, the last sentence was sarcastic or I meant the opposite of what I just said, which might be sarcasm…if we apply semantics? In any event, back to boobs! (BTW, how the hell do atomic orbitals exist in the sun with all that plasma? #1000000000Kelvin)

With the knowledge that elemental line spectrums are simply the result of electron decays in specific circumstances and there is a **possible** overlap between electron degradations in the sun and about hydrogen gas on Earth, what happens when you see a great pair of BOOBS? Well, as a result of the force to your ocular cavity, and the eyeballs there in, your head will twist to maintain focus on these boobs. All of which, is why fusion releases energy…albeit by matter degradation. (Stay focused people…the wow is coming.)

Let's imagine you have a girlfriend and she sees your head twist to ogle a perfect pair of boobs. Of course, you'll get a slap. Now, let's imagine you don't have a girlfriend and these perfect boobs cause your whole body to undergo a complete trajectory adjustment. This complete trajectory adjustment will result in a slight degradation to the rubber on the souls of your shoes…like a cartoonish tire screech. And that ladies and gentlemen, is the **FIRST** release of energy associated with fusion…if you imagine yourself as an electron seeing two perfect protons. The **SECOND** release of energy associated with fusion is when you CRASH into all the other dudes that also ATTRACTED to these perfect boobs. And the **THIRD** release of energy associated with fusion is when you crash into all those negative thermal energetic quanta hanging around these perfect boobs. Unfortunately, there is a caveat.

In my book, *Thinking Outside My Gated Community*, I made the postulate that all fusion is cold, which at the time made quite a bit a sense since I had just postulated that atomic orbitals coral negative thermal energetic quanta. Unfortunately, as is with most of my postulates, it's relative. Or in simpler terms, there is a spectrum of fusion events as it relates to cold and hot. (BTW, I should be more careful about making 'absolute' statements, i.e. it's very unscientific.) In any event, let me try and explain the relativity of this fusion spectrum.

In deep-dark-cold positive space, there's not a lot of negative thermal energetic quanta. Therefore, the "third" step of energy release is minimized. But, when fusion occurs in the presence of negative thermal energetic quanta, this will result in an enhanced "third" step of energy release. Or in simpler terms, the colder the environment around the fusion event, the colder the fusion event will be UNTIL the net energy release is **negative**. Conversely, the hotter the environment around the fusion event, the hotter the fusion event will be UNTIL the net energy release is **positive**.

In conclusion, fusion releases energy. Butt, this release of energy is the result of electrons degrading as they crash into negative thermal energetic quanta, which means the ENVIROMENT determines if fusion is HOT or COLD. All of which, is somehow related to boobs…I think. I get confused when boobs appear in front of me. In any event, the larger the fusion event, i.e. the larger the forged atom, the more the environment will **amplify** if it is a HOT or COLD fusion event.

Chapter 10: Auto-Chiralization

Some time ago, in the most reputable scientific magazine on this planet, *Nature*, some researchers reported that auto-chiralization occurs in some very specific reactions. Fortunately, they did NOT have the knowledge of stereo or electronic isotopes, because they might have used this as justification for auto-chiralization. (BTW, I bet you thought I was going to use that justification...didn't you?) In any event, this is going to be a very short chapter. Therefore, let me be honest with you about something completely unrelated...just for a moment.

When professors start using my books to teach, I'll be rolling in the money because I've made a conscious effort to include postulates from biology, chemistry, physics, astrophysics, and several other scientific sub-genres in most of my books. And even though this might seem greedy, which I am, it will also begin to break down the boarders between the sciences. Or simpler terms, collaborations will tots make your research better. Also, just imagine the brainiac kids that will be forged when most scientists start marrying across scientific kingdoms. In any event, you can rest assured that I'll only spend a couple million a year on strippers...usually in the form of free college tuition for them or their kids. And now, back to this transition chapter.

The spin of the star suggests the type of chiral plasma within. And since we know that CERTAIN spin electrons are stabilized by CERTAIN spin plasma, it is not outside the realm of logic that ONE type of chirality exists or is propagated around ONE type of star, which is spinning uniquely.

In conclusion, the Milky Way quanta degraded to produce star energetic quanta. And since the components of the Milky Way quanta were attracted to each other, opposite spins, it shouldn't be outside the realm of consciousness that opposite spin stars were produce. And since star quanta are also composed of smaller energetic quanta, it shouldn't be a surprise that a star with an abundance of ONE type of spinning energy will spin a certain direction. All of which means, stars with certain spins will release energy with a certain spin, which will stabilize atomic orbitals with certain electron spins…yada, yada, yada…auto-chiralization is a function of the sun's directional spin.

Chapter 11: Catalytic Cells

Even though we claim to know a lot about cells, we don't know shit. Mostly, because we refuse to factor in the shit coming from other cells. And as stanky as that may sound, multicellular organisms are based upon putting their garbage out on different days…metaphorically speaking of course. All of which, means there MUST be a rhythm to cellular division.

If we take cells absolutely back to their fundamental basis, catalysts surrounded by lipid bilayers, then cells are somewhat understandable: When conditions favor catalysis, the cell will divide. And when we view cells as catalytic entities, then the **variable** amount of time they spend at some stages in mitosis is the result of a LIMITING reaction. For example, DIFFERENT types of molecules surge through your blood stream at DIFFERENT times in your day. Therefore, if EACH cell has a different limiting reaction to make it FINISH mitosis, then different organs will repair at different times. Therefore, the current understanding of the mitosis timeline, is simply an AVERAGE mitosis timeline. (BTW, have you ever wondered why people with certain diets tend to develop certain types of cancers?) And now, the FACTOR that determines molecular transport in and out of cells.

In as much as humans are mostly water, the movement of ions in and out of cells is an extraordinary method to control catalysis. For example, a protein will have a different conformation in a 5 molar solution of sodium as compared to a 10 molar solution of sodium. (FYI, it is important to note that I said "sodium" and NOT sodium-chloride. Cells do an amazing job of selectively absorbing SPECIFIC ions, which will drastically modulate the conformation of proteins.) Therefore, keep in mind that UNIQUE cells have UNIQUE affinities to ions and respond differently to slight variations in ionic concentrations in and around cells, which may be a factor in different cells undergoing mitosis at different times.

In conclusion, the fundamental basis of cells is nothing more than RESTRAINED catalysis, which dictates cellular division. Or in other terms, the mitosis timeline that is currently being taught is an AVERAGE of all the UNIQUE limiting reactions, which are dependent on molecules the body circulates at different times. (Have you ever wondered how ALL the cells in OLD people are about the same relative age? #SuckItTelomereTheory) Or in evolutionary terms, the rhythmic mitosis of different cells in a multicellular organism will be a great way to maintain homeostasis. Or in other words, NON-rhythmic mitosis will INCREASE the number of genetic-copying errors since EVERY cell needs an uninterrupted energy supply when going through the most important parts of mitosis. All of which, leads back to cells paying attention to their neighbor's shit…as weird as that sounds.

Chapter 12: Histones

I've always wondered how environmental factors modulate the genome from one generation to the next. But, it didn't make any sense until I started thinking about histones: The most terrific suitcase unknown to humanity. Unfortunately, before I get to all the "environmental factors" crap, I have to do the unthinkable: Talk about sex in a culture that has grown from puritanical roots. (#BagagePun) In any event, let me start at a relative beginning.

For those of you who don't know this, sperm are streamlined delivery devices, which is where the term "streamlined" came from…I think. But, an ovum, not located in the butt, is a FULL cell with extra room for extra things. (#JunkInTheTrunk) And as such, it wouldn't be surprising that an OVUM, with 90 years of life to live, has extra histones lying around. (FYI, telomere DNA will break off if there aren't any histones to wrap around. #SuckItAgainTelomereTheory.) All of which, brings us to a very romantic setting: Two glasses of wine, a candle, and a defective condom…oops?

With minuscule amounts of disgusting stuff, unless you're into poop and pee sex, a sperm combines with an ovum…crash test dummy style. And via my previous postulates, the amino acids in the sperm's streamlined rock-hard body reignites the ovum's mitosis process with a slight twist: Holliday Junctions, which RANDOMLY occur between the paternal and maternal chromosomes. Actually, that was a lie. If Holliday Junctions occurred **randomly**, then Holliday Junctions might result in a tall and short chromosome. (Damn, maybe that's how we ended up with 23 chromosomes…in an evolutionary sense?) In any event, for whatever reason, Holliday Junctions are NOW relatively specific. All of which means, there is a **protein** that binds to the HISTONES that exist about similar regions in the maternal and paternal chromosomes. I wonder what EXCESS proteins might be laying around, which have an unusual affinity to histones? How about histones?

Hypothetically, ALL alleles are frozen, i.e. wrapped around histones, when the Holliday Junctions occur. But, what directs the LOCATON of Holliday Junctions? Well, as any good geneticist knows, the ONLY factor that determines gene expression is a promotor region...regardless of how they might be hidden about the histone polymer. (FYI that was sarcasm.) In any event, let's take a moment to postulate something sillier than normal.

Previously, I postulated that XX-females were the first species and that XY-males were a common occurrence until the limiting "amino acid switch" evolved to focus reproduction about more fruitful environments. Therefore, and this is very important, excess histones in the XX-ovum are probably a factor in the formation of the Holliday Junctions. Butt, it gets even weirder.

After the Holliday Junctions are complete, the 23-pairs of chromosomes are wrapped around a mixture of histones: 51% mom and 49% dad. But, now the cell must undergo mitosis, i.e. copy all 23-pairs of chromosomes. Or in annoying scientific terms, after the Holliday Junctions, there are at least 46 alleles that code for HISTONES, half maternal and half paternal, and NOW THE CELL MUST MAKE HISTONES FOR 23-**NEW**-PAIRS OF CHROMSOMES, i.e. mitosis. What decides which histone allele will get expressed? Butt, as always, the complexity doesn't stop here.

Since each human cell does NOT express genes from each chromosome, the percentage of paternal to maternal histones will vary after differentiation. Or in other words, different organs will have different percentages of paternal or maternal histones. All of which, is a PERFECT method as to how offspring develop certain paternal/maternal characteristics/diseases. Or in simpler terms, a tissue that expresses a lot of the mother's histones, will express more of the mother's alleles because the allele's preferential arrangement about the mother's histones. Or in simplest terms, the mother's histones probably hide the promoter regions to the father's alleles simply by variable wrapping of the DNA or histone polymerization. (Too bad skin color isn't determined like cats' fur color. #FRECKLES) Thankfully, this brings us back to the first quandary: How do environmental factors modulate the genome from one generation to the next?

Histones are hydrophilic on the outside and hydrophobic on the inside, which allows the intercalation of different amounts and types of fats into the histones, which will modulate the histone's three-dimensional structure. And as a result of the histone's different three-dimensional structure, the DNA wrap will be different and histone polymerization will be different. All of which, will modulate the efficacy of promotors/genes.

In conclusion, if you're one of those nerds that have already sequenced your genome and your parent's genome, don't FREAK-OUT! Until scientists can determine which cells express which histones and/or how much of each histone they express, REMEMBER that science still doesn't understand ALL the complexity. The best thing you can do is eat healthy and supplement your diet with branch chained amino acids…since your body can't make them. And finally, no matter how LONG your telomeres might be, they will indefinitely break off if there aren't enough histones to wrap around. (BTW, it is possible that certain fats over long periods of time will modulated the histone polymer and weaken the association of the histones responsible for stabilizing the telomere…but that just me thinking out loud.)

Chapter 13: MSSFs

More people suffer from depression in America than any other country, which is probably because of the food we eat. Also, diabetics are more likely to suffer from depression, which is VERY interesting since more people suffer from diabetes in America than any other country. (BTW, diabetics also have nerve problems, which I'll talk about later.) In any event, based upon my previous postulate that neurons run on microfluidic dynamics and NOT electricity, I think Myelin Sheaths are being overlooked as factors in several disease states. In any event, before we get to the little Myelin Sheath man, let's do a quick review.

At the end of a nerve, pressure pushes synaptic vesicles of Serotonin out into the synaptic cleft. Then, Serotonin binds to the transmembrane ion channels, which opens them up and allows water/ions to flush into the second synaptic bulb. Therefore, SSRIs, via the microfluidic theory, are increasing the FLUSH power of each nerve impulse by increasing the number of Serotonin molecules in between the 'communicating' neurons. Or in simpler terms, inhibiting the re-absorption of the Serotonin will increase the flush power of each nerve impulse. But before we go any further, let's review how the SSRI therapy came into existence.

First, scientists discovered how to blend up tissue samples and then isolate proteins via density centrifugation. Next, these transmembrane proteins were placed into vesicles. Finally, tons of small molecules were tested, with Serotonin, to identify inhibitors. (FYI, this process is tots interesting and really complex because scientists use substrates, catalysts, and pharmacophores to identify the inhibitors.)

Unfortunately, OLD scientists had NO REASON to believe that Myelin Sheaths had SSRI sensitive transmembrane ion channels, which is why scientists did NOT separate the Myelin Sheaths from their tissue samples. Therefore, there's no way to definitively state that Myelin Sheaths DO NOT have SSRI sensitive transmembrane ion channels. All of which, leads to my postulate: Myelin Sheaths & synaptic bulbs have the same transmembrane ion channels.

I realize that you are hesitant to believe that Myelin Sheaths have the same transmembrane ion channels, because I'm crazy, but let me ask you three questions: In very LONG neurons with multiple Myelin Sheaths, HOW do transmembrane ion channels get in-between Myelin Sheaths? Second, since proteins need to be replaced from time-to-time, where do the old proteins go? And finally, do you have good credit and $99? (If so, then you can get a new Hyundai!)

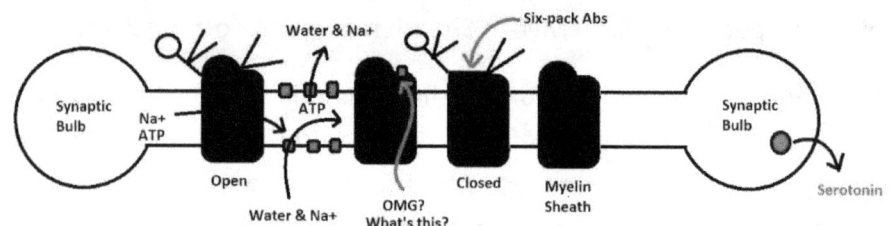

Figure 26: Six-pack Abs...after liposuction?

Hopefully you can see, Myelin Sheath's FATASS is blocking the diffusion of transmembrane ion channels FROM the synaptic bulb to BETWEEN the Myelin Sheaths. Or in simpler terms, OMG, what's this? (Look at Figure 26 again if you're confused.) Or in other terms, OMG, the RED transmembrane ion channel in the Myelin Sheath, which, as I'm about to show you, does the SAMETHING as the transmembrane ion channels in the neuronal tube. Butt first, turn to the next page.

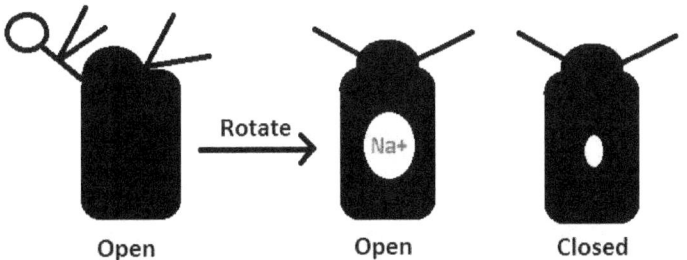

Figure 27: I see London, I see France, I see BUTTHOLES!

As you can see, I'm using the butthole of this little man to describe the neuronal tube. Now, let's imagine a positive neuronal flush breaches this little man's butthole. (#eww) Well, since the function of a Myelin Sheath is to increase flow by decreasing back-flow, the Myelin Sheath constricts the little man's butthole, i.e. the neuronal tube. All of which, should make you wonder: What provides the FORCE to constrict this little man's butthole? (BTW, did you think my shitty neuronal explanation was going to get any cleaner?) In any event, here is my postulate.

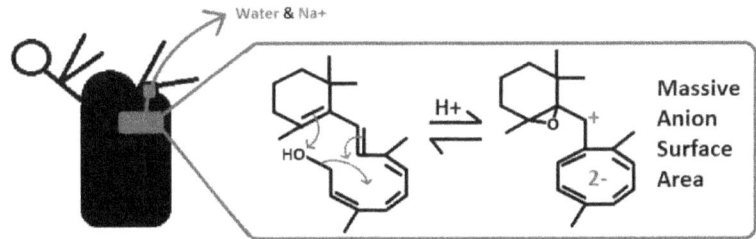

Figure 28: Pericyclic Vitamin A Bladder...eww?

Within this little lipophilic man, there is a hydrophilic cytosol bladder region. And as a result of the little man peeing out water and sodium, via his tiny transmembrane ion channel, the equilibrium of this reaction is shifted towards the MASSIVE ANION SURFACE AREA molecule. And, as you probably already surmised, these MASSIVE ANION SURFACE AREA molecules dissolve into the lipid bilayer and surge towards the POSITIVE ions raging through the little man's butthole, which causes the little man's sphincter to constrict, i.e. liposuction. (FYI, for those of you who aren't familiar with the hard soft behavior of cations, i.e. H+ & Na+, it is possible to manipulate one concentration in order effect the other concentration. Or in simpler terms, the anions that were associated with the Na+ will suck up the H+ without the cytosol becoming "acidic"...thus effecting the equilibrium.)

Now, I realize that most people are firmly applying the brakes to this shitty explanation. Butt, let me remind you that pericyclic reactions are NOT uncommon in the human body. For example, a fat soluble precursor to Vitamin D undergoes a pericyclic reaction to produce lipophilic Vitamin D in the skin, which is why we hate Nigers. (Wow, I think that racist statement just woke-up about a thousand scientists.) In any event, even though that racist statement is ILLOGICAL, scientifically speaking, it is NOT illogical that the FORCE behind the neuronal tube constriction is the result of the **negative** pericyclic vitamin A derivatives surging towards the **cations** within the neuronal tube. (Chemistry and Biology fucked and had a baby named BINGO!) In any event, let's take a step back and add in a bit more logic.

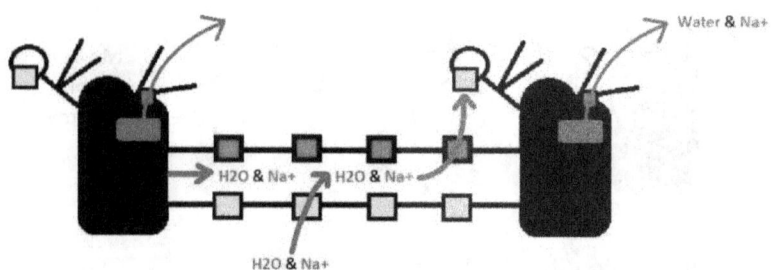

Figure 29: Disgusting & Sexual?

First, the neuronal flush pushes water and sodium down the neuronal tube. Second, the positive cations chelate the ORANGE transmembrane proteins, which opens them up to allow MORE water/sodium to flush into the neuronal tube. Third, after the propagation of the neurological flush, the RED transmembrane ion channels pump the water and sodium out. Fourth, the little man Myelin Sheath sucks up the water and sodium via his ORANGE transmembrane ion channels, which modulates the cytosol equilibrium and relaxes the Myelin Sheath, i.e. forms tons of NON-anionic & NON-pericyclic vitamin A molecules. And finally, after the neuronal tube stops squirting water and sodium into the little Myelin Sheath man's face, the little Myelin Sheath man can return to a resting state by piss-out all the excess water and sodium, which prepares the little Myelin Sheath man for the next neuronal flush, i.e. forms tons of anionic pericyclic vitamin A molecules. (FYI, the lymph system sucks up the excess water and sodium so that it doesn't bother the Myelin Sheath anymore.) Or in simpler terms, Myelin Sheaths **contain** and **supply** BOTH types of transmembrane ion channels, red and orange, to the neuronal tube in the spaces between the little Myelin Sheath men. (Tots beautiful…A?) All of which, leads to the true 'zinger' of this system.

If you haven't noticed, the body continually absorbs and excretes material on a regular basis. Now, in an evolutionary sense, this is to prevent cellular damage by the relaxation of isotopes, i.e. maintain a non-acidic environment. Therefore, and this is the most important point of the chapter, the microfluidic theory of nerves allows the body to maintain this "continual flush". Or in simpler terms, vascularization about nerves endings will provide ENERGY to the neuron ALONG with water and sodium, which gets flush down the neuronal tube, out around the neuronal tube, into the Myelin Sheath, out of the Myelin Sheath, and into the Lymphatic system. Or in the SIMPLEST terms, the OLD theory of nerves has STAGNANT water and sodium hanging out around neurons indefinitely, which goes against the fundamental framework of the human body, i.e. continual flushing. And now, a few more odd things before the conclusion.

The job of the Myelin Sheaths is to prevent backflow of the neuronal flush, which requires energy. And even though Myelin Sheaths are fatty-esk cells, it is ODD that diabetics suffer depression at higher rates and will have diabetic nerve pain if they live with high sugar levels. Or in simpler terms, low sugar means depression and high sugar means diabetic nerve pain. All of which, can be linked to Myelin Sheaths, i.e. too much squeezing, extra pain; too little squeezing, depression. (FYI, the complexity of sugar/fat ratios and the energetics of Myelin Sheaths will be discussed in a later book.)

In conclusion, I could go into a long explanation of how Vitamin A rich spinach has helped my IBS and depression, but I've already talked about way too much shit in this chapter...I even talked shit. (BTW, we're all human and deserve the same respect. Sorry about the racist illogical wake-up comment.) And finally, these things ARE logical: **A)** Myelin Sheaths must contain the same type of transmembrane ion channels because they supply transmembrane ion channels to the neuronal tube between Myelin Sheaths; **B)** Vitamin A is fat soluble and Myelin Sheaths are fatty-esk cells. **C)** The formation of a HUGE anionic hydrophobic vitamin A pericyclic entities will allow Myelin Sheath's lipid bilayer to surge towards the positive charges within the neuronal tube, i.e. constrict the neuronal tube; **D)** It's NOT surprising that anti-depressants, which modulate transmembrane ion channels, might ALSO modulate the transmembrane ion channels in fat-cells and Myelin Sheaths; **E)** The human body is based upon continual flushing. (BTW, I'll talk about how Myelin Sheaths deliver transmembrane ion channels to the neuronal tube in a later book...hopefully. Also, MSSFs stands for Myelin Sheath Selective Facilitators.)

Chapter 14: Krebs' Pickle

In the last chapter, I made an impassioned plea to the **logic** of continual flushing as it relates to the nervous system and human body. All of which, brings me to another oddity: Based upon FLOW, with regards to neurological tubes, the molecules in the Krebs cycle are NOT energetically equivalent. As always, let me try and explain.

In as much as everybody hates the oil companies, supposedly, the oil companies know one thing better than anybody else, i.e. VISCOSITIES. And in a much as viscosity, in a rudimentary sense, is based upon molecular properties, oil companies know better than to try and PUMP grease through pipelines. Or in simpler terms, it takes a lot of energy to pump grease in comparison to crude oil. All of which means, the molecules in the Krebs cycle are not energetically equivalent. Unfortunately, since the nervous system is an aqueous based system, I need to move past a rudimentary understanding of viscosity.

In an aqueous system, viscosity is often related to the number of dissolved ions. But, in the sense of an organic system, this also includes the amount of small molecules, e.g. ATP. All of which results in this question: If most Myelin Sheaths are NOT highly vascularized, how do Myelin Sheaths obtain the energy they need to function? Well, ATP is a hydrophilic entity and Myelin sheaths are hydrophobic, so that's not a logical option. Next, even though high sugar concentrations, in a diabetic state, often causes high sugar levels in the interstitial fluid about the Myelin Sheaths, this is not a viable option since most of us are not diabetic...yet? All of which, points to the fact that most Myelin Sheaths get their energy from processing fatty acids, which is VERY convenient since Myelin Sheaths are fatty-esk cells. All of which, brings us back to viscosity in aqueous solutions.

Based upon the microfluidic theory of neurological impulses, the movement of information down the neuronal tube is the result of cationic concentrations. And for those of you who don't know this, enhance cationic concentrations will modulate the solubility of small fatty acids. Or in simpler terms, cationic neuronal flushes will leach out and **move** the SMALLEST fatty acids down the neuronal pipeline. Therefore, if Myelin Sheaths have excess small fatty acids in their lipid bilayer, then more than likely these small fatty acids will diffuse into the neuronal tube's lipid bilayer and eventually into the neuronal flush. All of which, brings us to the Krebs cycle.

The Krebs cycle has one function: Covert the energy stored in long chain fatty acids into ATP. And as a result of this, continually shorter fatty acids are produced. (For those who have studied the Krebs cycle, you're probably annoyed by the fact that the fatty acids are always covalently bound to a substrate in the Krebs cycle. I feel you annoyance, but NOT all Krebs cycles are the same, i.e. different cells have different proteins and different proteins may or may not hydrolyze under different circumstances.) In any event, based upon the non-vascularization of most Myelin Sheaths and their hydrophobic existence, I postulate that the Krebs cycle within Myelin Sheaths allows for the abnormal adjustment of fatty acids within the cell and its membrane. All of which, brings us back to the concept of energetically different fatty acids.

For a moment, let's imagine two girls with two different perfumes: One girl has a perfume with a 14-carbon fatty acid chain and the second has a perfume with 5-carbon fatty acid chain. Now, imagine which perfume diffuses faster. Or in simpler terms, the chemical bonds do NOT have more energy, they simply APPEAR to have more energy based upon their diffusion rate. Or in the simplest terms, smaller fatty acids have more "energy" because they diffuse faster in comparison to larger fatty acids. All of which, makes smaller fatty acids the perfect communication molecule between Myelin Sheaths. But first, prostaglandins.

Prostaglandins are these MASSIVE hydrophobic molecules that have caused most humans to lament the atrociousness of nature...from time to time. And as such, it is not outside the realm of consciousness that smaller fatty acids may also be used as communication between Myelin Sheaths. (Short paragraph, but you get my point...hopefully.)

So let me recap my diffuse theory so far: **A)** Myelin Sheath's Krebs cycle produces an abnormally amount of small fatty acids, which diffuses into the Myelin Sheath's cellular membrane and then into the neuronal tube's lipid bilayer; **B)** The enhance concentration of cations in the neuronal flush leaches out the smaller fatty acids and transports them further down the neuronal tube; **C)** In addition to these smaller fatty acids re-enforcing the neuronal tube, i.e. keeping ions from crossing, the smaller fatty acids diffuse into the next Myelin Sheath; **D)** As a result of the "more energetic" smaller fatty acids, each subsequent Myelin Sheath becomes better tuned to the activity of the previous Myelin Sheath. Or in simple terms, Myelin Sheaths are communicating via small fatty acids, which is part of the physiological learning process.

For a moment, let's imagine there's ABSOLUTELY no physiological response by the first Myelin Sheath when a nerve cell undergoes an unsuccessful neuronal flush. (Oh the absolute horror of a non-adaptive system that can't adjust to variations in external stimuli!) Ok, now let's imagine that repetitive unsuccessful neuronal flushes cause the first Myelin Sheath to flex its anal-existence. At some point, the Myelin Sheath neuronal sphincter is going to get tired. And what do tired muscles do when they get tired? (Release Lactic Acid) Therefore, overtime, the first neuronal sphincter will inform the second neuronal sphincter of its shitty existence...via the release of small fatty acids.

In conclusion, babies are fucking lazy because it takes a lot of energy and shit to condition neurological pathways. And since Myelin Sheaths are fatty-esk cells and are connected by a tube of flushing cations, which can ostensibly dissolve smaller fatty acids, it is not outside the realm of consciousness that Myelin Sheaths communicate via modulated Krebs cycles, which produce smaller fatty acids. In addition to that, it wouldn't be surprising if these smaller fatty acids also maintain the neuronal tube's integrity against ions. And finally, based upon the rate of diffusion, smaller fatty acids are more energetic than longer fatty acids. (BTW, have you ever wonder if the Lactic Acid from tired muscles ENHANCES the activity of Myelin Sheaths, which communicates to your brain that your muscles hurt?)

Chapter 15: Structure Equals Function

For those of you who don't know this, a precursor to Vitamin D undergoes a pericyclic reaction in the skin, as a result of negative photons, to form Vitamin D, which is fat soluble. Also, chlorophyll's aromatic heme molecule focuses photons towards magnesium, which sits slightly below the aromatic heme. All of which, sets the stage for pericyclic Vitamin A in the eyes…at least in my brain.

The current theory of eyesight is that a SINGLE photon causes the isomerization of vitamin A's double bonds, from trans to cis, in the eye. Then, the cis-vitamin A associates with and opens a transmembrane ion channel, which creates a neuronal flush. Now, I could go on a rant about how the OLD theory of 'sight' is based upon quantum mechanics and the ABSURD idea of an itsy-bitsy-tiny-whiny photon causing the excitation of a HUGE electron to allow the double bond to isomerize, but instead, I'll simply say this: It takes a lot of negative thermal energetic quanta to cause the isomerization of a double bond. Therefore, on the premise that photons do NOT degrade into a **trillion** negative thermal energetic quanta, which would have the POWER to distort a HUGE negative atomic orbital, let's VIEW sight as a MULTIPLE photon endeavor, i.e. via Quanta Dynamics.

When sight is viewed as a MULTIPLE photon endeavor, as PER THE CONCEPT of the fucking lenses in your eyeballs, you may begin to wonder HOW the other components of the eye AID in the 'FOCUSING' of photons. Well, we already know that plants "generally" focus photons via electromagnetic proteins and an aromatic heme, but eyeballs focus SPECIFIC photons. Therefore, there must be a method by which DIFFERENT transmembrane ion channels respond to DIFFERENT photons. Fortunately, I have just the example: Pericyclic vitamin A.

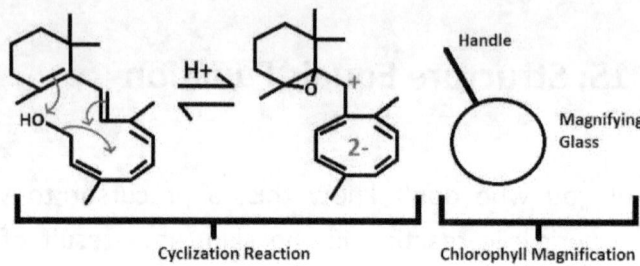

Figure 30: Photo-isomerizations are weird.

As you can surmise from the last chapter, I'm under the assumption that vitamin A is like vitamin D and can undergo a pericyclic reaction...either by photo-isomerization or protein facilitation. Now the only difference between vitamin A & D, is that vitamin A produces an aromatic structure and vitamin D produces a complex conjugated hydrocarbon structure. Or in simpler terms, vitamin D's pericyclic reaction is NOT reversible, in terms of undergraduate consciousness, and vitamin A's reactions exists as an equilibrium. In any event, let's use pericyclic vitamin A's focusing ability to better understand the selective detection of certain photons.

As discussed in the *"Catalysis Obsession"* chapter, aromatic rings ARE magnetically active. (FYI, this is super interesting because magnetically chiral aromatic rings are probably the method by which birds navigate via the Earth's magnetosphere.) And on top of all that, I've postulated that aromatic rings ALSO focus plasma. Therefore, to understand this theory, you just need to ask yourself one question: What is energetically between magnetic energetic quanta and plasma energetic quanta? (Hopefully, you guessed photons.) All of which, leads to the PINNACLE of my theory: Aromatic pericyclic vitamin A focuses photons onto the atomic orbitals of a specific amino acid, which causes the photons to degrade into thermal energetic quanta; All these negative thermal energetic quanta distort a specific amino acid atomic orbital, which causes a conformational change in the amino acid...then the protein...then the transmembrane ion channel, i.e. initiates a neuronal flush. Or in simpler terms, it's like burning an ant with a magnifying glass, except the ant is an amino acid that undergoes a conformational change, which results in the opening of a transmembrane ion channel. But, before we go any further, let's talk blind fish. (Have you ever wondered if blind fish taste different?)

If you've ever taken a day off from the insane American grind, gotten high, and watched the Nature Channel, then you've probably stumbled upon a show talking about blind cave fish. Yada yada yada, it's not outside the realm of consciousness that since the "eyeball system" is based upon aromatic rings, which ALSO focuses negative plasma, that blind cave fish see heat. But enough about blind fish, let's get back to meat eaters.

With the pinnacle of my ocular theory being a LESS THAN obvious extension of people wearing glasses, but on a photonic level, let me clarify a few specifics:

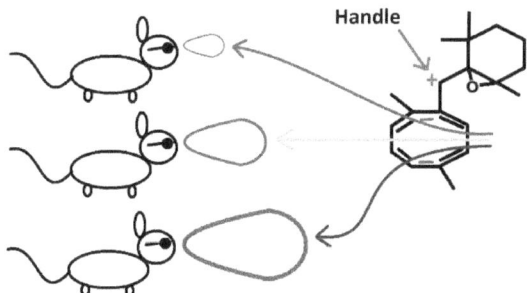

Figure 31: Three blind mice.

Let's imagine there are three legally blind mice that simply need healthcare to see again. (FYI, this example is really stupid, but bear with me.) But, each mouse has a different "atomic orbital" that needs to be distorted in order to activate a transmembrane ion channel. Therefore, the doctor will have to take the focusing device, i.e. pericyclic aromatic vitamin A, by the HANDLE and focus the light on each different atomic orbital. Or in simpler terms, the movement of the pericyclic aromatic vitamin A to different regions on the transmembrane ion channel will be a factor in the detection of specific wavelengths. (BTW, have you ever wondered how birds know which berries to eat based upon their genetic code? Well, it's simple. The genetic code is responsible for the transmembrane ion channels, which determines the placement of the pericyclic aromatic vitamin A. Or in simple terms, birds are more sensitive to specific wavelengths, which appear brighter to them based upon their inherited protein motifs.) Unfortunately, specific wavelength detection is insanely more complex than just the placement of pericyclic aromatic vitamin A, i.e. it includes physics, chemistry, and Quanta Dynamics. (Have you noticed that I shamelessly plug my new realm of science more often than I should?) In any event, not to be overwhelming, let me simply talk about the chemistry aspect...for the time being. But first, evolution.

I realize that God created everything in 7 days and our sun was **_never_** white, i.e. contained more RED and ORANGE photons. But, if God hypothetically created humanity via a projection from the order within atomic particles, and the patternistic way they react, then **evolution** is the ONLY way birds would have been able to survive on Moon-Earth and then on Earth. Or in simpler terms, bird proteins had to TUNE to different wavelength concentrations under a yellow star. (If I'm wrong, then I'll burn in hell...or the Texas sun, whichever is hotter.) In any event, back to heathenistic chemistry aspect.

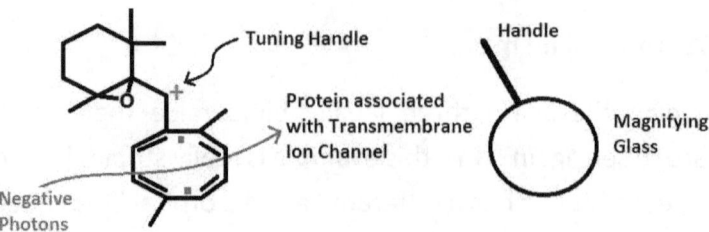

Figure 32: "Tune in Tokyo! Tune in Tokyo!"[7]

After about eight years of graduate school, I finally got it through my head that aromatic rings can be tuned based upon their substituents. In particular, Chapter 6 of this book. All of which, I've applied to my postulate of pericyclic aromatic vitamin A. Or scientific terms, the MASSIVE steric hindrance the carbocation in Figure 31 could result in the carbocation associating with an amino acid residue without forming a covalent bond. And furthermore, it is not outside the realm of science that this carbocation is vibrating between the functional groups of two amino acids. All of which, will have TREMENDOUS tuning ramifications on the pericyclic aromatic vitamin A molecule, which I'll discuss later. ("Later" implies this discussion will occur in one of my future books...if you were wondering.)

In conclusion, I have set the stage with something beautiful...if you can see it. Quite simply, based upon the fact that conjugated double bonds can undergo pericyclic reactions, it is not outside the realm of intuition that vitamin A forms an aromatic pericyclic anionic molecule. And since aromatic rings have the ability to focus energetic quanta, it is highly probably that eyeballs are based upon FOCUSING energy...even at the molecular level. All of which, when extrapolated to evolution, birds, and blind fishes, excluding blind mice, makes things seem less complicated. Unfortunately though, as will be the case in my future books, the application of physics to this system will make things seem uncontrollably complicated. And finally, even though I didn't mention it in this chapter, I previously mentioned that different photons degrade into different thermal energetic quanta and that different atomic orbitals are distorted differently by this different energy, which provides another level of selective photonic detection by transmembrane ion channels. (BTW, have you ever wondered if the conformational change in the transmembrane ion channel results in the donation of a proton, which pushes the vitamin A equilibrium back in the direction of NON-pericyclic vitamin A, i.e. a METHOD by which a transmembrane ion channel doesn't get burnt...like an ant under a magnifying glass on a sunny day?)

Chapter 16: Rates

After a lot of thought, I finally figured out the problem with vaccines as it relates to autism…and you're NOT going to like this. Quite simply, the only factor that can NOT be accounted for is the HEALTH of the child, which determines the RATE at which the child's body reacts to vaccines. Or in simpler terms, the STRESS associated with vaccines is 10X worse in children that are malnourished or undernourished. All of which, leads to an increase in autistic children amongst minorities.

Even though statistics is NOT America's strong suit…anymore, it is what drives science. And even though everybody thinks it's a big conspiracy, healthcare professionals actually care about helping people and kids. But, who is the most likely blame someone else for their kid's autism? (The poor, uneducated, and malnourished?) Therefore, as the scientists and doctors LOOK at the statistics of autism and vaccines, there is a VERY VERY small correlation between the **rate** of vaccine doses and autism. But, since doctors cannot "empirically" assign HEALTH to a child to be included in the vaccine data, they can NOT see the correlation. (FYI, malnourished children don't always have high temperatures.) Therefore, malnourished or unhealthy children undergo MORE stress as it relates to the **rate** of vaccines, which causes autism.

In conclusion, if this postulate is correct, there are two ways to go about fixing this problem. First, you can DECREASE the rate of vaccines in small children. Second, you can require vitamin shots be given before or with all vaccine doses. (We could stop feeding our kids shit, but that's not an option in America?) In any event, I guess it's NOT a surprise that as the RATE of home cooked meals decreases, there is an increase in the RATE of autism. Or in less vague terms, as the rate of breast feeding decreases, the rate of autism increases. (FYI, if you want a baby, you should understand that this means your nipples will be fucking **raw** for at least a year.)

Chapter 17: PiHKAL with THC and Nicotine

It is disgusting that a large portion of society abuses drugs. And it is even more disgusting that this drug abuse is costing America trillions of dollars every year. Unfortunately, not everyone has the analytical hootspa to comprehend what they are feeling as it relates to nicotine...or other drugs.

For some time, I was on the cigarettes gravy train. But, bronchitis always wanted to hangout, so I decided to quit. At first, I used nicotine gum and patches to help break my habit, but those things did NOT provide the SAME relief as cigarettes. (FYI, nicotine is an analgesic like aspirin and ibuprofen...or so the current theory goes.) All of which, I didn't think about until sometime later...after Breaking Bad.

For a similarly long time, I enjoy weed in graduate school. Ironically, it provided me with a bit of relaxation and didn't cause much of a hangover, in comparison to alcohol. But, it wasn't until after synthesizing THC from a legal drug (CBD) in my bedroom and adding it into my pancakes, that I came across the same conundrum as with nicotine: Why did THC pancakes create a body buzz, which was totally different than smoking weed? (BTW, from my experience, THC and CBD combat the effects of ruffies. So, take some CBD or smoke a joint before you go out and you'll be less likely to get raped...hopefully. Actually, depending on what you smoke, the paranoia might just steer you clear of all the freaky people?) All of which, lead me back to my scientific training.

For those of you who are not familiar with flash vacuum pyrolysis, I pity you. (Just kidding.) In any event, highly reactive intermediates are created under HEAT/vacuum and then captured in super cold glass jars. Now, being that these highly reactive intermediates are, how you say "highly reactive", they are usually reacted with something to form a more stable compound, which can be isolated, purified, and analyzed. All of which, allows for the logical deduction of the reactive intermediate, via know chemistry, i.e. booooring! Any who, yada yada yada, based upon the knowledge that oral nicotine does not sate pain the same as burnt tobacco, i.e. cigarettes, there seems to be another molecule in cigarette smoke that has a SUPER analgesic property. Unfortunately, scientists have not been able to find this molecule…because they're not looking for it?

For those of you who don't claim to be cognizant with regards to research on nicotine abuse, let me help: There is a genetic correlation between people who abuse cigarettes and people who don't abuse cigarettes. Or in simpler terms, alcoholics tend to run in the family and cigarettoholics also tend to run in the family. Butt, what is it about familiar genetics that facilitates the abuse of cigarettes? Is it a sensitive receptor, as the current research implies? Me say no. (Bet you didn't see that one coming.) Any who, let me indulge you in MY train of thought…since that's what you're paying my blind trust for. (How many billions is it now?)

The one factor that researchers have been missing is: Lung proteins. Quite simply, ANY reactive intermediate formed by burning tobacco will be trapped by the lung proteins, i.e. emulsifying proteins. And even though we currently THINK that nicotine is the analgesic culprit, here's my postulate: There is a molecule within the cellular membrane of the tobacco leaves that undergoes thermal rearrangement to form a reactive intermediate, which reacts with the emulsification proteins to form a morphine-esk analgesic.

It is actually quite simple, if you think about it. A reactive intermediate from burnt tobacco reacts with a lung protein to form a morphine-esk structure. All of which, is the REASON why oral nicotine isn't the greatest method to helping people quit smoking, i.e. fuck nicotine…there's something else going on in burnt tobacco.

Regardless of how you feel about tobacco verses weed, or your limited education as to why ONE is legal and the other isn't, the psychotropic behavior of weed is probably similar to the analgesic behavior of tobacco, i.e. burnt tobacco decreases muscle pain and burnt weed decreases the pain of time. (Ah weed…the soft drifting of consciousness away from the shit reality of American incarceration.)

Having said all this, it would be civically irresponsible if I didn't add in a bit of honesty into this pseudo-civics/chemistry quandary. Quite simply, drugs have been associated with creative types: artists, musicians, and theorists. (Einstein used cocaine…good for him!) But, based upon my brain function postulates as well as from personal experience, I need to come clean. Yes, drugs allow for the expansion of thought, but I propose that the pain associated with the cession of drugs inspires the mind to look for something soothing: maybe a tune, maybe a picture, maybe an idea, or maybe 12 steps. It is also my assumption, that the greatest athletes use the pain of continual exertion to imagine the feeling of success. All of which means, mental conditioning is also the key to success in athletics.

In conclusion, WHAT we obsess about when we are uncomfortable is what drives us to be the people we desire to be. Unfortunately, these soothing thoughts are typically not rationale and often lack TRUE scientific data and/or educated historical perspective. (Please, for the love of God, just look up how and why weed was made illegal in America OR just try some pot pancakes. They're nothing like smoking weed.) But for some of us, with minds trained by science, we can identify certain differences and theorize why these differences occur. (I just wish America had Equal Education such that everybody has a chance to see something scientifically...even if it is just the smell of your farts as a function of the food you eat.) And finally, nicotine pills and patches SUCK at helping people to quit smoking, which is probably because there is another reactive molecule in burnt tobacco that is forging a morphine-esk structure, which will be nerded upon in the next chapter.

Chapter 18: Diels Alder

If you ever want to give an organic chemistry professor a boner, ask him or her about Diels Alder. But, in as much as I love talking about old-boners, I need to bring up the absence of boners: Cardiovascular disease. Now for many of you, this affliction has never crossed your mind. But for people with a cigarette addiction, this thought is never far from their consciousness. All of which, brings us back to flash vacuum pyrolysis.

Cooking...yummy, right? The mixing of exotic spices, sodium bicarbonate, and/or vinegar makes your mouth water. All of which, leads to one important question: Where does the water go? (FYI, you swallow it.) Unfortunately, LUNGS can't swallow.

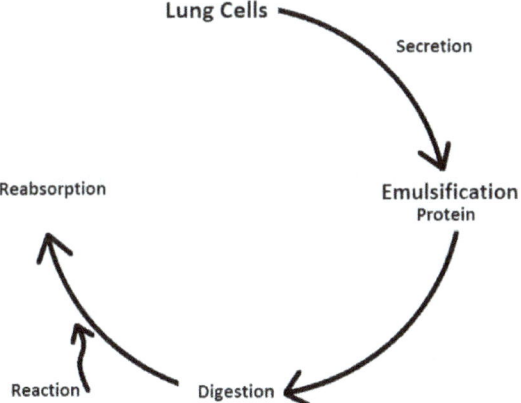

Figure 33: Making drugs in your lungs...sweet!

As you can see in this figure, nothing lasts forever...especially the proteins in your lungs when they are being bombarded with reactive intermediates, which are forged in the crucible of a cigarette. All of which, gives a FAST viable pathway to the absorption of morphine-esk protein motifs into the circulatory system...and that patented high you pay for with every box.

Now, in terms of Quanta Dynamics, it is NO surprise that copious amounts of negative thermal energetic quanta can distort hydrogen bonds. But for the sake of being anal, let's see what happens in a fag.

Figure 34: Burning Cigarettes.

As you can see, tobacco leaves have cells and cells have fatty acids. When fatty acids are exposed to copious amounts of negative thermal energetic quanta, i.e. a burning cigarette, they lose hydrogens to form double bonds, which we call unsaturated fatty acids. Then, these unsaturated fatty acids can undergo a Diels Alder with a fragment of a lung protein to form heterocycles, which mimic the structure of morphine. All of which, is the reason why Diels Alder is a star...or a tobacco leaf? In any event, there is ONE thing about all this that should make your head hurt: There are a TON more fatty acids in tobacco leaves in comparison to nicotine! Hmmm, am I onto something here?

Have you ever had decaffeinated coffee? Also, have you ever had a de-nicotined cigarette? If so, then you know "light" cigarettes are LESS addictive? I mean, I have NEVER seen an old lady smoke two packs of menthol lights and then say "They're not addictive" with her mechanical voice box…or have I? Also, have you ever seen a chemo-patient hooked up to an IV outside a hospital smoking a cigarette because nicotine patches just don't alleviate the pain as well as old fashion morphine cigarettes? Yeah, something else is going on in with regards to this cigarette addition and I think it is the formation of morhine-esk protein motifs via flash vacuum pyrolysis, which diffuses into the blood and sates muscular pain.

With all that said, based upon the postulate that tobacco burning yields a morphine-esk protein adduct that is active to the PNS, i.e. relieves muscular pain, BUT nicotine can cross the blood brain barrier, it is NOT surprising that researchers opted to blame nicotine for ALL the addictive behavior. Much in the same way, researchers opted to blame THC for the psychotropic behavior because it can cross the blood brain barrier. But again, the digestion of nicotine/THC does NOT have the same physiological effect as burnt tobacco/weed. All of which, leads to the uncomfortable conclusion that PNS nerve innervation can result in a psychotropic effect. Or in simpler terms, THIS IS THE WHOLE REASON why certain food combinations can be addictive…without any small molecule crossing the blood brain barrier. Or in the simplest terms, substances don't have to cross the blood brain barrier to innervate the CNS. (#McDonaldsFood)

In conclusion, our ancestors discovered tobacco because it made them feel good, i.e. relieved pain. But since then, tobacco has become a monetary beast. Therefore, it is not surprising that research about tobacco's "additive behavior" has been blocked…countless times. Or in simpler terms, are researchers rewarded with HUGE monetary grants to look at the addictive properties of nicotine when tobacco is addictive via another process, which is just ANOTHER way for the tobacco industry keep its product on the market? Or in the simplest terms, if EVERY scientist agrees that tobacco is addictive because of nicotine, but the addiction is the result of burnt tobacco fatty acids reacting with emulsifying lung proteins to form a morphine-esk structure, then people will never be able to break the tobacco habit. In any event, the digestion of nicotine/THC does NOT have the same physiological effect as burnt tobacco/weed. All of which, puts new light on the PNS's influence on the CNS. And finally, the euphoric effect of cigarette smoke diminishes for a given amount of time based upon the availability of emulsification protein fragments in the lungs to form morphine-esk structures, which correlates back to one's genetic make-up.

Chapter 19: Soooo?

So why do I write short "theoretical" science books? Well, I used to do it so that people would listen to my logic about social decency. Unfortunately, there aren't many people that have the diverse scientific vocabulary to read my books…even though they're really short. (Either that or the grammatical/spelling errors are just too annoying. #Trolling) Therefore, I'll save my social commentary for my non-scientific books. But, in honor of science being about questions, I'd like to end this book with one last question: We could be better people…if we tried?

References

1. *Spaceballs*. Mel Brooks. MGM. 1987. (Cover Quote.)
2. *Futurama*. Cohen, D.X.; Goening, M., 20th Century Fox Television. 1999.
3. Organic Chemistry 2nd Edition. Paula Yurkanis Bruice. Prentice Hall. New Jersey. 1998.
4. For any X-ray or NMR data pertaining to the acclaimed Compound X, please contact Professor Stephen C. Bergmeier at 277 Clippinger Hall, Ohio University, Athens, Ohio. (bergmeis@ohio.edu.)
5. Barraclough, P.; Young, D.W.; Ferrige, A.G.; Lindon, J.C., The H-1 Nuclear Magnetic Resonance of 6-substituted Bicyclo-[3.1.0]Hex-2-Enes. *Journal of Chemical Society-Perkin Transactions 2* **1982**, (6), 651-656.
6. *CRC Handbook of Chemistry and Physics*. Florida: CRC Press, 1994, 74th edition.
7. *Girls just want to have fun*. Metter A., New World Pictures. 1985.

www.ingramcontent.com/pod-product-compliance
Lightning Source LLC
Chambersburg PA
CBHW061206180526
45170CB00002B/980